Little Book of Big Eco Actions
Seven Generation Sustainability

Dr Jonathan Frost
With GPT 4.0

Polycrisis

The simultaneous occurrence of several catastrophic events.

Action

The fact or process of doing something, typically to achieve an aim.

The Seventh Generation Principle

Decisions we make today should result in a sustainable world seven generations into the future.

ISBN: 9798858067580
Imprint: Independently published

Table of Contents

Introduction

It's now or never.

If you're not telling your clients or employer to act in alignment with the solutions in this guide, then what's the point? We don't need more box-ticking eco-consultants intentionally or unintentionally supporting business as usual. The world is on fire in between the floods and people are beginning to wake up.
As experts in the field we've presented the facts, submitted petitions, attended global summits, lobbied government, glued ourselves to immovable objects and demonstrated in city centres.
Not much has changed.
We don't have time to beat around the bush. We need to politely and firmly tell our clients, employers, and connections to take meaningful action today. Another certification form filled in won't cut it. We must use every method of persuasion in our toolbox to make action happen. Remember personal growth occurs at the edge of our comfort zone.

How to use this guide

Have a conversation with your client.
You:
"Hi there! I understand that you're concerned about sustainability and that you want to take some action that makes a difference.

Tell me two or three aspects of sustainability that particularly concern you."
Client:
"Well, I'm particularly saddened by pictures of deforestation and waste plastics everywhere."
You:
"I understand. Let's look at a few things that you can do that align with your core values, and those of your business."
[Choose a few actions from the book and show them to the client]
"Which of these options do you find most attractive?"
Client:
"This one and this one."
You:
"Great! I'll put together a short and long-term plan so we can get the ball rolling right away."
Client
"Thanks. I'm really keen to get going with this."

Go through the eco-action plan that you give to your employer, colleagues and clients.
Match every action in your plan to the closest related element in the lists in this guide (great acceleration, UN SDGs or societal behaviours).
Reevaluate or remove any items in your plan that don't align with suggestions in this guide.
Check links on the **Green Elephant** website for references and successful projects.

Fix our problems and predicaments

The current polycrisis has elements related to the great acceleration, the United Nations' 17 sustainability development goals and global societal structure. Some problems appear in more than one of these three categories and the grouping below is arbitrary.

Focus on solutions

We have identified most of the problems we face. This guide focuses on solutions and immediacy-
10% problems stated
70% solutions offered
20% additional context
Climate emergency action plan checklist

Project viability

Before we waste our time and energy let's ask a few questions:

Is it solving the right problem?
Is it scientifically viable? Does it obey the known laws of physics?
Is it technologically viable? Is there a demonstration project working today?
Is it scalable? Are there resources available for everybody in the world to do this?
Is it resources viable? Is there a sustainable supply of minerals available?

Is it land a viable? Is there enough space available without taking away from a forest or agricultural land?
Is it politically viable? Can it be made acceptable to the general population?

References

New useful information appears continuously. Please visit **green-elephant.uk** for the current bibliography and website links.

Feedback

Please let me know of any factual errors.
Please let me know of significant omissions.
Please tell me your recommended updates with references.

Problems and predicaments

The Great Acceleration

"The Great Acceleration" is a term used to describe the rapid advancement of various measures of human activity since the mid-20th century. It essentially captures the dramatic, exponential growth of virtually every measure of human activity, from population to technology to the economy.

1. **Population growth**: The human population has rapidly expanded from around 2.5 billion in the 1950s to well over 7 billion today.

2. **Urbanization**: More people live in cities today than in rural areas, which has major impacts on land use and habitat loss.

3. **Industrial output**: We've seen a dramatic increase in industrial production and the global economy, leading to more consumption and waste.

4. **Energy consumption**: There's been a massive increase in the use of fossil fuels, leading to greenhouse gas emissions and climate change.

5. **Freshwater use**: Water use has surged, causing issues with water scarcity in many regions.

6. **Fertilizer consumption**: The use of synthetic fertilizers has increased exponentially, affecting ecosystems and leading to phenomena like dead zones in the ocean.

7. **Land system change**: Widespread conversion of land for agriculture, industry, and urban development has led to significant habitat loss and species extinction.

8. **Deforestation**: Large swathes of forests, especially in tropical regions, have been cleared for agriculture and other uses.

9. **Biodiversity loss**: We're experiencing a mass extinction event due to habitat loss, pollution, climate change, and other factors.

10. **Climate change**: The increase in greenhouse gas emissions from human activities is causing global warming and climate change.

11. **Ocean acidification**: The oceans are absorbing large amounts of CO2, leading to acidification and damage to marine ecosystems.

12. **Plastic pollution**: We've seen a dramatic rise in the production and disposal of plastic, much of which ends up in the oceans.

13. **Air pollution**: Increased industrial activity, energy production, and transportation have led to significant air pollution.

14. **Nitrogen and phosphorus cycles**: Human activities, especially agriculture, have significantly altered these critical ecological cycles.

15. **Chemical pollution**: The production and use of many synthetic chemicals, including pesticides, has had wide-ranging impacts on human and environmental health.

16. **Transport:** Personal and commercial transport by land, sea and air has increased exponentially since 1960

17. **Data transfer and the Internet:** High bandwidth video makes up more than 80% of current data traffic on the internet and uses energy and resource hungry infrastructure.

18. **Exponential technological development**: Artificial intelligence, robotics and pattern recognition impact business systems and working practice.

19. **Soil degradation:** Intensive and mechanized agriculture destroys in decades soil structure that has taken millennia to create.

United Nations' 17 Sustainability Goals not covered specifically in the great acceleration list

1. Inequality
2. Food
3. Personal health
4. Disease
5. Education
6. Peace, justice and strong institutions
7. Globalisation
8. Transport

Societal Behaviour

Over-consumption of energy and resources
Misinformation, censorship and mass media control
Business as usual and the military industrial complex
Excessive bureaucracy
Social cohesion
Global debt
Global financial system
Giant Tech Monopolies and FANG

Solutions – Great Acceleration

Population growth

1. **Promote Family Planning**: Access to family planning services, contraceptives, and education about reproductive health can empower people to make informed decisions about the size of their families.

2. **Improve Education**: Access to quality education, particularly for girls, can have significant impacts on family size, as more educated women tend to have fewer children and have them later in life.

3. **Gender Equality**: Advancing gender equality can have a significant effect on population growth. Empowered women often choose to have fewer children, contributing to slower population growth.

4. **Reduce Child Mortality**: Reducing child mortality rates can decrease the perceived need for large families, thereby slowing population growth.

5. **Economic Development**: Economic development and reducing poverty can lead to lower birth rates as people don't need to rely on having more children for economic security.

6. **Social Security Systems**: Implementing social security systems can reduce the need for large families, as parents don't have to rely on their children for support in their old age.

7. **Urban Planning**: Adequate urban planning and investment in infrastructure can help accommodate increasing populations in sustainable ways.

8. **Invest in Agriculture**: Increasing agricultural efficiency can help feed a growing population without overtaxing natural resources.

9. **Promote Sustainable Practices**: Promoting sustainable consumption and production practices can help ensure resources are used efficiently.

10. **Climate Action**: Addressing climate change can help ensure the long-term habitability of the planet, despite a growing population.

While managing excessive global population growth presents challenges, it's crucial to remember that population control measures must respect human rights and dignity. The focus should be on voluntary, informed choices, and creating conditions that favour sustainable population growth.

Urbanization

1. **Urban Infill**: By developing vacant or underused land within existing urban areas, cities can become denser and make better use of their infrastructure, reducing the pressure for suburban growth.

2. **Public Transportation Improvements**: Enhancing public transportation can make living in the city more appealing and reduce the demand for suburban housing. Additionally, expanding public transportation to suburban areas can lessen the environmental impact of commuting.

3. **Mixed-Use Zoning**: Encouraging mixed-use developments can make cities more attractive places to live and work, reducing the desire for suburban living. These developments combine residential, commercial, and industrial uses, making amenities more accessible.

4. **Green Spaces**: Creating and preserving green spaces in urban areas can make city living more attractive, potentially reducing the desire for suburban homes with yards.

5. **Affordable Housing Initiatives**: By ensuring affordable housing is available in urban areas, cities can keep a diverse range of residents and reduce the push towards the suburbs.

6. **Urban Revitalization Projects**: Revitalizing derelict or neglected areas can attract residents and businesses back into the city, reducing suburbanization.

7. **Sustainable Building Practices**: Encouraging sustainable building practices in cities can reduce the

environmental impact of urban living and make cities more appealing places to live.

8. **Tax Incentives**: Offering tax incentives for urban living or businesses operating in urban areas can help keep people and businesses in the city.

9. **Education and Public Services**: Investing in city schools and public services can make urban living more attractive for families who might otherwise move to the suburbs.

10. **Smart Growth Policies**: Implementing smart growth policies, which aim to create sustainable, inclusive, and vibrant communities, can counter the negative effects of suburbanization.

Each of these solutions has its strengths and weaknesses and must be adapted to fit the local context. Moreover, implementing these solutions requires a balanced approach, considering the economic, environmental, and social aspects of urban development.

Industrial output

Addressing the need for genuine sustainability in manufacturing involves developing efficient, innovative, and responsible strategies. Here are the top 15 solutions to enhance sustainability in the sector:

1. **Resource Efficiency**: Manufacturing processes should be designed to utilize resources (material and energy) as efficiently as possible. This includes reducing waste and optimizing the use of raw materials.

2. **Renewable Energy**: Factories should shift towards using renewable energy sources, such as wind, solar, and hydropower. This reduces the carbon footprint associated with manufacturing.

3. **Circular Economy**: This involves designing products for longevity, repairability, and recyclability, thus encouraging the reuse and recycling of products and materials.

4. **Lean Manufacturing**: Adopting lean principles can improve efficiency, reduce waste, and decrease energy use.

5. **Sustainable Procurement**: Selecting suppliers who follow sustainable practices can help to extend sustainability throughout the supply chain.

6. **Product Life-cycle Assessment**: This is a method to assess the environmental impacts associated with all stages of a product's life. It helps to identify where improvements can be made.

7. **3D Printing**: Also known as additive manufacturing, this can help to reduce waste by using only the materials needed to build a product.

8. **Green Buildings**: Manufacturing facilities can be designed or retrofitted to be more sustainable, including energy-efficient lighting and HVAC, water-saving fixtures, and more.

9. **Water Management**: Implement strategies to reduce, recycle and reuse water in manufacturing processes.

10. **Training and Education**: Workers should be trained on sustainable practices and the importance of sustainability in manufacturing.

11. **Waste Management**: Factories should implement strong waste management procedures, including reducing waste, recycling, and composting.

12. **Energy-Efficient Equipment**: Upgrade to more energy-efficient machinery and equipment to reduce energy consumption.

13. **Eco-Design**: The design process should consider the environmental impacts of a product from its inception.

14. **Eco-Labelling**: This can help consumers make more sustainable choices, driving demand for sustainably produced goods.

15. **Industry Collaboration**: Sharing best practices and knowledge among companies can lead to industry-wide improvements. Collaboration can also lead to joint research and development efforts to innovate more sustainable processes and products.

The challenge lies in implementing these solutions on a large scale. This will require commitment from all levels of the organization, government support, and consumer awareness and demand for more sustainable products.

Energy consumption

Addressing the problem of excessive global energy consumption requires strategies that both reduce energy demand and transition energy production to sustainable sources. Here are some potential solutions:

1. **Energy Efficiency**: Implementing energy-efficient technologies and practices in buildings, industry, and transportation can significantly reduce energy demand.

2. **Renewable Energy Transition**: Transitioning to renewable sources like wind, solar, and hydropower can provide energy with fewer environmental impacts.

3. **Sustainable Urban Design**: Designing cities to minimize energy use, for example by prioritizing public transportation, cycling, and walking, can reduce energy demand.

4. **Smart Grid Technologies**: Modernizing the electrical grid with smart technologies can optimize energy distribution and reduce waste.

5. **Demand Response Programs**: These programs incentivize consumers to reduce demand at peak times, helping to balance the grid and reduce overall energy use.

6. **Green Building Standards**: Implementing green building standards can make buildings more energy-efficient, reducing the energy required for heating, cooling, and lighting.

7. **Education and Awareness**: Increasing awareness about the importance of energy conservation can encourage

individuals and organizations to reduce their energy use.

8. **Vehicle Fuel Efficiency Standards**: Implementing and enforcing fuel efficiency standards for vehicles can reduce energy consumption in the transportation sector.

9. **Carbon Pricing**: Implementing a carbon tax or cap-and-trade system can incentivize energy efficiency and the use of clean energy sources.

10. **Decentralized Energy Systems**: Local, decentralized energy systems can reduce transmission losses and encourage the use of local renewable energy sources.

11. **Public Transportation**: Investing in reliable and efficient public transportation can reduce reliance on energy-intensive private vehicles.

12. **Energy-Efficient Appliances**: Promoting the use of energy-efficient appliances can significantly reduce household energy consumption.

13. **Research and Development**: Investing in research and development can lead to new technologies and strategies for reducing energy consumption.

14. **Regulation and Policy**: Implementing policies that mandate energy efficiency and renewable energy can drive systemic changes in energy consumption.

15. **Industrial Process Improvements**: Upgrading equipment and improving processes in industries can significantly increase energy efficiency and reduce overall energy use.

Each of these strategies can contribute to reducing global energy consumption. However, it's important to recognize that significant reductions in energy consumption will require systemic changes to how energy is produced and used, rather than solely relying on individual actions.

Freshwater use

1. **Water Conservation**: Encourage and educate individuals, businesses, and industries to use water more efficiently. This could involve simple measures like fixing leaks, as well as more significant changes, such as adopting water-efficient technology.

2. **Desalination**: Expand the use of desalination technology, where seawater is converted into fresh water. As this process can be energy-intensive, efforts should be made to utilize renewable energy sources.

3. **Wastewater Recycling**: Treat and reuse wastewater for non-drinking purposes such as irrigation, industrial processes, or flushing toilets.

4. **Rainwater Harvesting**: Collect and store rainwater for later use. This can be a particularly effective solution in areas with regular rainfall.

5. **Improved Irrigation**: Promote the use of more efficient irrigation methods in agriculture, such as drip irrigation, to reduce water waste.

6. **Protecting and Restoring Watersheds**: Protect and rehabilitate natural water sources like rivers, wetlands, and aquifers to maintain their ability to provide clean water.

7. **Infrastructure Improvement**: Update and maintain water infrastructure to prevent water loss through leaks and inefficiencies.

8. **Virtual Water Trade**: Implement policies to trade water-intensive goods from water-rich to water-poor

regions, reducing the pressure on water resources in arid areas.

9. **Water Pricing**: Implement water pricing that reflects the true cost of water provision and encourages water conservation.

10. **Community-Led Management**: Encourage community-led water management to ensure the needs of the local community are met and that water sources are responsibly managed.

Each of these solutions offers a way to address the global lack of fresh water, but their feasibility and effectiveness will depend on local conditions. Moreover, implementing these solutions requires cooperation and commitment at all levels, from individuals to international organizations.

Fertilizer consumption, Nitrogen and phosphorus cycles

Excessive use of fertilizers can lead to a host of environmental issues, including water pollution, soil degradation, air pollution, and harm to wildlife. Here are 10 potential solutions to address these problems:

1. **Precision Agriculture**: Using technology like GPS, remote sensing, and data analysis, farmers can determine the precise amount of fertilizer their crops need and prevent overuse.

2. **Use of Organic Fertilizers**: Organic fertilizers, such as compost and manure, are usually slower releasing and less likely to cause nutrient runoff compared to synthetic fertilizers. They also help improve soil health and structure over time.

3. **Plant Buffer Strips**: Planting buffer strips along the edges of fields can help to absorb excess nutrients before they reach water bodies. These strips can be made up of grasses, trees, or other plants.

4. **Crop Rotation**: Different crops have different nutrient needs. By rotating crops, farmers can help balance the nutrients in the soil, reducing the need for additional fertilizers.

5. **Soil Testing**: Regular soil testing can help farmers understand the nutrient levels in their soil and only apply the required amount of fertilizers.

6. **Cover Crops**: Planting cover crops can help absorb excess nutrients left in the field after the main crop has

been harvested. This can prevent nutrient runoff and also help improve soil health.

7. **Integrated Nutrient Management**: This approach combines the use of organic and inorganic fertilizers to maintain soil fertility and productivity, reduce costs, and minimize environmental impact.

8. **Education and Training**: Farmers and gardeners should be educated and trained about the potential harm of excessive fertilizer use and how to use fertilizers responsibly.

9. **Regulations and Policies**: Government regulations and policies can limit the use of certain types of fertilizers or incentivize best management practices.

10. **Promoting Sustainable Farming Practices**: This can involve various practices that aim to protect and enhance the natural resources upon which agriculture depends, including reducing synthetic fertilizer use, conservation tillage, and more.

By adopting these solutions, we can significantly reduce the problems generated by the excessive use of fertilizers.

Land system change

Land-use change, excluding deforestation, can have profound impacts on ecosystems, including habitat loss, soil degradation, and climate change, among others. Here are the top ten solutions to mitigate the negative impacts of such changes:

1. **Urban Planning**: Smart city planning, such as zoning, can help balance the needs of development and conservation. Green spaces within urban areas can also help maintain biodiversity.

2. **Land Restoration and Rehabilitation**: Restoration of degraded land and soil can improve their productivity and ecosystem services, helping to reduce the need for new agricultural land.

3. **Sustainable Agriculture Practices**: Techniques such as agroforestry, permaculture, and organic farming can minimize the environmental impact and maintain long-term productivity.

4. **Mixed-Use Development**: Mixing residential, commercial, and recreational land uses can reduce the need for land change by creating self-sufficient communities where people live, work, and recreate.

5. **Regenerative Agriculture**: This agricultural system promotes the regeneration of the soil, increasing its biodiversity and improving its health, which can increase its resilience to land use changes.

6. **Land Use Policies**: Implementing policies that encourage sustainable land use and discourage

wasteful or harmful practices can be a powerful tool for managing land use change.

7. **Buffer Zones**: These can be used to separate different land use types, such as agricultural and natural lands, to prevent negative impacts from one to the other.

8. **Education and Awareness**: Promoting understanding of the value of ecosystems and the impacts of land use changes can lead to more responsible decision-making at all levels.

9. **Conservation Easements and Land Trusts**: These legal agreements allow landowners to protect their land from certain types of development, preserving it for future generations.

10. **Incentives for Sustainable Practices**: Government subsidies or tax breaks can incentivize landowners to maintain or switch to sustainable land use practices.

Each of these strategies can help to address the negative impacts of land use change. Often, the most effective approach will involve a combination of these solutions, tailored to the specific context and needs of the area.

Deforestation

Deforestation has serious environmental impacts, including loss of biodiversity, disruptions to the water cycle, soil erosion, and carbon emissions, contributing to climate change. Here are the top ten solutions to mitigate the negative impacts of deforestation:

1. **Reforestation and Afforestation**: Planting trees in areas where forests have been cut down, or creating new forests in areas where there weren't any before, can help to restore some of the benefits provided by forests.

2. **Sustainable Forest Management**: This involves the responsible use of forest resources, including selective logging, replanting, and other techniques that preserve the health of the forest ecosystem.

3. **Legal Protections**: Enacting and enforcing laws to protect existing forests and restrict illegal logging can be very effective at preventing deforestation.

4. **Community-Based Forest Management**: Empowering local communities to manage their own forests can provide them with a sustainable source of income and incentivize them to preserve the forest.

5. **Payment for Ecosystem Services**: This system compensates individuals or communities for preserving forests, recognizing the value that these ecosystems provide.

6. **Agricultural Improvements**: Implementing sustainable farming practices, such as agroforestry and

permaculture, can reduce the need to clear more land for agriculture.

7. **Consumer Awareness**: Educating consumers about the impacts of deforestation and encouraging them to buy products made with sustainable materials can help reduce demand for products that contribute to deforestation.

8. **Reduction of Forest Fires**: Implementing policies and practices to prevent and control forest fires can help preserve forests.

9. **Green Certifications**: Certifications for products sourced from sustainably managed forests can provide an economic incentive for forest conservation.

10. **International Cooperation**: Since deforestation is a global problem, international agreements and cooperation can play a key role in addressing it.

These strategies can help reduce deforestation and mitigate its impacts, but they require commitment and cooperation from individuals, businesses, and governments.

Biodiversity loss

1. **Protected Areas**: Expand and effectively manage protected areas, including terrestrial and marine reserves, to safeguard key biodiversity zones.

2. **Sustainable Agriculture**: Promote farming practices that protect and enhance biodiversity, such as agroforestry and organic farming.

3. **Control Invasive Species**: Implement measures to prevent, control, and eradicate invasive species that threaten native biodiversity.

4. **Ecosystem Restoration**: Engage in large-scale efforts to restore ecosystems that have been degraded or destroyed, including forests, wetlands, and coral reefs.

5. **Legislation and Enforcement**: Develop and enforce laws to protect endangered species and habitats, and to control illegal wildlife trade.

6. **Education and Awareness**: Raise public awareness about the importance of biodiversity and the threats it faces, to generate support for conservation efforts.

7. **Sustainable Fishing and Hunting**: Regulate and manage fishing and hunting to ensure it is sustainable and does not lead to population declines or ecosystem imbalances.

8. **Combat Climate Change**: Addressing climate change is crucial for preserving biodiversity, as warming temperatures and changing precipitation patterns can disrupt ecosystems.

9. **Green Infrastructure**: Incorporate green spaces and natural elements into urban planning and development to support urban biodiversity.

10. **Finance and Incentives**: Provide funding and economic incentives for biodiversity conservation, such as payments for ecosystem services and ecotourism.

Biodiversity loss is a global crisis, threatening ecosystem stability, human food security, and our ability to respond to new challenges such as climate change. These solutions offer a way to address this crisis, but they require concerted, coordinated efforts at the global, national, and local levels.

Climate change

Addressing the issue of increased greenhouse gas (GHG) emissions is crucial for mitigating the impacts of climate change. Here are 15 potential solutions:

1. **Renewable Energy Transition**: Replace fossil fuels with renewable energy sources such as wind, solar, and hydropower.

2. **Energy Efficiency**: Improve energy efficiency in buildings, transportation, and industry to reduce energy demand.

3. **Reforestation**: Plant trees to absorb CO_2 from the atmosphere, while also preventing deforestation and promoting sustainable forestry practices.

4. **Sustainable Agriculture**: Adopt sustainable farming practices, including organic farming, permaculture, and regenerative agriculture to reduce emissions from the agriculture sector.

5. **Carbon Capture and Storage**: Develop and deploy technologies to capture CO_2 at the source (e.g., power plants) and store it underground.

6. **Electrification of Transport**: Replace gasoline and diesel-powered vehicles with electric ones, and improve public transportation and infrastructure for biking and walking.

7. **Green Building Standards**: Adopt green building standards to make new buildings more energy-efficient, and retrofit existing buildings.

8. **Circular Economy**: Develop and promote systems for reducing, reusing, and recycling materials to minimize waste and emissions from material production.

9. **Education and Awareness**: Increase public understanding and awareness about the causes and impacts of climate change to build support for climate action.

10. **Dietary Changes**: Encourage diets lower in meat and dairy, which are resource-intensive to produce, and promote locally sourced food to reduce transportation emissions.

11. **Policy Measures**: Implement carbon pricing (like a carbon tax or cap-and-trade system), renewable energy standards, and other policies to incentivize emission reductions.

12. **Green Finance**: Increase investment in clean technologies and companies, and divest from fossil fuels.

13. **Conservation and Restoration of Natural Ecosystems**: Protect and restore carbon-rich ecosystems like peatlands, mangroves, and rainforests.

14. **Decentralized Energy Systems**: Promote local, decentralized energy systems to reduce transmission losses and promote resilience.

15. **International Cooperation**: Work together on a global scale to reduce emissions, as climate change is a global problem that requires coordinated global action.

These solutions require systemic change and collective action. The technology and knowledge to significantly reduce GHG emissions already exist; what is needed is the political will and societal commitment to apply them on a broad scale.

Ocean acidification

Here are the top 15 solutions to mitigate ocean acidification that don't directly involve reducing CO_2 emissions:

1. **Ocean Alkalinity Enhancement**: Adding alkaline substances like crushed limestone to the ocean can neutralize some of the acidification.

2. **Protection and Restoration of Blue Carbon Ecosystems**: Coastal ecosystems like mangroves, seagrasses, and salt marshes can absorb and store large amounts of CO_2, helping to limit acidification.

3. **Reduce Other Ocean Stressors**: By limiting pollution, overfishing, and destructive fishing practices, we can help marine ecosystems maintain resilience against acidification.

4. **Promote Sustainable Seafood Choices**: Encouraging the consumption of sustainably harvested seafood can help reduce pressure on species most vulnerable to ocean acidification.

5. **Educate Public and Stakeholders**: Increasing awareness about the impacts of ocean acidification can lead to changes in behaviour and greater support for mitigation efforts.

6. **Enhance Marine Protected Areas (MPAs)**: Expanding and enforcing MPAs can help protect vulnerable species and habitats from additional stressors, increasing their resilience to acidification.

7. **Shellfish Aquaculture Practices**: Some farming methods can reduce acidification locally, for example, shellfish farming can absorb some of the excess CO2.

8. **Selective Breeding and Assisted Evolution**: Breeding species that are more tolerant of acidification can help sustain populations under changing conditions.

9. **Implement Sustainable Tourism Practices**: Ensuring tourism activities don't exacerbate acidification impacts can help preserve marine environments.

10. **Promote Climate Resilient Fisheries**: Implementing management strategies that increase the resilience of fisheries to climate change can help support communities and ecosystems facing acidification.

11. **Develop Acidification Early Warning Systems**: These systems can provide critical information to communities and industries that depend on marine resources.

12. **Ocean Liming**: This involves the application of fine limestone particles to the ocean surface, which can neutralize acidifying waters.

13. **Support Indigenous and Local Knowledge**: Indigenous and local communities often have unique knowledge and adaptive strategies that can contribute to addressing ocean acidification.

14. **Funding Research**: Additional research can help improve our understanding of ocean acidification, develop effective mitigation strategies, and monitor progress.

15. **International Cooperation**: Given the global nature of ocean acidification, international cooperation is crucial to implement effective solutions.

These strategies can all contribute to mitigating the effects of ocean acidification, by enhancing the resilience of marine ecosystems, promoting sustainable practices, and increasing our understanding of this complex issue.

Plastic pollution

Here are 15 solutions that address plastic pollution post-production, focusing on practical actions that individuals, communities, companies, and governments can take:

1. **Recycling Programs**: Implement widespread recycling programs that accept a variety of plastics and educate citizens about what can and can't be recycled.

2. **Clean-Up Campaigns**: Organize and participate in clean-up efforts in local communities, beaches, parks, and waterways to manually remove existing plastic pollution.

3. **Waste Management Infrastructure Development**: Improve and expand waste management systems to ensure plastics are properly disposed of or recycled, reducing the chance they end up in the environment.

4. **Education and Public Awareness**: Develop campaigns to educate the public about the importance of proper plastic disposal and the impacts of plastic pollution.

5. **Plastic Collection Programs**: Create programs that encourage the collection of plastic waste from households, businesses, and public spaces.

6. **Extended Producer Responsibility (EPR)**: Implement policies that hold manufacturers responsible for the end life of their products, encouraging them to develop take-back or recycling programs.

7. **Incentivize Plastic Return**: Implement deposit-refund systems for plastic containers to encourage return and recycling.

8. **Zero Waste Programs**: Encourage the adoption of zero waste practices at the community or city level, which can include initiatives like composting and widespread recycling.

9. **Bioremediation**: Encourage research into and use of organisms like bacteria or fungi that can degrade or break down plastics in the environment.

10. **Pyrolysis and Plastic-to-Fuel**: Implement technologies that convert plastic waste into fuel or other useful products, making waste plastic valuable and reducing the likelihood it ends up in the environment.

11. **Upcycling**: Encourage and promote the practice of creatively reusing plastic waste to create new products of higher value.

12. **Landfill Management**: Improve landfill management to prevent plastic waste from escaping into the environment.

13. **Adopting Circular Economy Principles**: Implement systems that keep materials in use for as long as possible, extract the maximum value from them, recover and regenerate products at the end of their service life.

14. **Promote Industrial Symbiosis**: Encourage industries to utilize the waste from one company as a resource for another where possible, effectively reducing the overall waste.

15. **Regulation of Microplastics**: Implement strict regulations around the use and disposal of

microplastics, particularly in industries such as
cosmetics.

Each of these solutions can help manage and reduce the
amount of plastic waste that ends up in our environment post-
production. They involve various stakeholders, from
consumers to waste management companies, businesses, and
policymakers.

Air pollution

Here are 15 solutions to address air pollution that do not directly involve reducing emissions:

1. **Green Urban Planning**: Urban greening, like planting trees and creating green roofs, can help absorb pollutants in the air.

2. **Promote Public Transportation, Biking, and Walking**: Encourage these transportation methods to decrease the number of vehicles on the road, indirectly reducing emissions.

3. **Indoor Air Purification**: Use air purifiers and regularly replace filters in homes and workplaces to remove pollutants.

4. **Waste Management**: Improve waste management systems to reduce open burning of waste, which can contribute to air pollution.

5. **Zoning Regulations**: Implement zoning in cities to separate industrial areas from residential areas to minimize human exposure to pollution.

6. **Education and Awareness Campaigns**: Enhance public awareness about air pollution, its causes, effects, and how individuals can help reduce it.

7. **Weather Forecasting and Warning Systems**: Develop systems to forecast pollution events and inform vulnerable populations to reduce exposure.

8. **Use of Clean Cooking Stoves**: In developing countries, traditional stoves often burn biomass and coal, which

contribute to indoor air pollution. Cleaner stoves can reduce this problem.

9. **Plant-Based Diets**: Encourage diets low in meat and high in organic, locally sourced foods to reduce the environmental impact of agriculture, which contributes to air pollution.

10. **Regular Maintenance of Vehicles**: Encourage regular vehicle maintenance to ensure they are running efficiently and emitting fewer pollutants.

11. **Wind Energy**: Utilizing wind energy can help dilute air pollutants and facilitate dispersion.

12. **Urban Design**: Design cities to improve air flow by considering building heights and street widths, which can help disperse air pollution.

13. **Promote Teleworking**: Encourage companies to allow remote work to reduce the number of commuters.

14. **Dust Management**: In arid regions or construction sites, manage dust to prevent it from becoming airborne and contributing to particulate pollution.

15. **Resilient Health Infrastructure**: Develop healthcare facilities and services prepared to handle the increased health problems associated with poor air quality.

Remember that while these strategies can help manage air pollution and its impacts, the most effective long-term solution is to reduce the emissions that cause air pollution in the first place. Nonetheless, these suggestions can help minimize the impacts of air pollution on human health and the environment.

Chemical pollution

1. **Regulatory Measures**: Implement and enforce strict environmental laws and regulations for industries, businesses, and individuals to control the release of pollutants into the environment.

2. **Promote Renewable Energy**: Encourage the use of renewable energy sources such as solar, wind, and hydro to reduce pollution from fossil fuels.

3. **Waste Management**: Improve waste management systems to reduce the amount of waste that ends up in landfills, water bodies, and the atmosphere.

4. **Cleaner Production Techniques**: Invest in cleaner technologies and production methods that minimize pollution and waste.

5. **Sustainable Transportation**: Promote the use of public transportation, cycling, walking, and electric vehicles to reduce air pollution from vehicles.

6. **Consumer Awareness and Education**: Educate consumers about the environmental impacts of their consumption habits and encourage them to choose eco-friendly products.

7. **Recycling and Upcycling**: Encourage recycling and upcycling to manage waste and reduce the need for new raw materials.

8. **Water Treatment**: Invest in water treatment infrastructure to clean polluted water and make it safe for human use and the environment.

9. **Carbon Capture and Storage**: Invest in technologies that capture and store carbon dioxide emissions to mitigate air pollution and climate change.

10. **Eco-friendly Agriculture**: Encourage eco-friendly farming practices that reduce pollution from agrochemicals, like organic farming and integrated pest management.

11. **Reforestation**: Reforestation efforts can absorb carbon dioxide, reduce air pollution, and prevent soil erosion.

12. **Plastic Alternatives**: Promote the development and use of alternatives to single-use plastic products, which are a major source of pollution.

13. **Restoration of Natural Habitats**: Restoring natural habitats can help absorb or mitigate the impacts of various forms of pollution.

14. **Community Initiatives**: Encourage community clean-up initiatives for local environments like beaches, parks, and neighbourhoods.

15. **International Cooperation**: Pollution doesn't respect national boundaries, so international cooperation is crucial for managing and reducing pollution on a global scale.

Implementing these solutions can significantly reduce global pollution and its impacts on human health and ecosystems. However, it requires collective action from governments, businesses, communities, and individuals alike.

Transport

Global transport, both personal and commercial, contributes significantly to carbon emissions, noise pollution, congestion, and air pollution. Here are 15 solutions that could help address these problems:

1. **Public Transportation**: Improve and expand public transportation systems to provide a more efficient alternative to personal car use.

2. **Active Transportation**: Encourage walking, cycling, and other forms of active transportation through urban planning and infrastructure development.

3. **Car Pooling and Sharing**: Promote carpooling and ride-sharing to reduce the number of vehicles on the road.

4. **Telecommuting**: Encourage companies to allow employees to work from home, reducing the need for daily commuting.

5. **Flexible Work Hours**: Allow flexible work schedules to reduce the number of people traveling during peak hours and alleviate congestion.

6. **Electric Vehicles**: Promote the use of electric cars, buses, and trucks, which produce less pollution than traditional vehicles.

7. **High-Speed Rail**: Develop high-speed rail networks to provide a fast and efficient alternative to air and car travel for longer distances.

8. **Efficient Logistics and Supply Chains**: Improve logistics to reduce the amount of travel required to

transport goods. This could include better route planning or consolidation of loads.

9. **Urban Planning**: Develop cities in a way that minimizes the need for transportation, for example by creating mixed-use neighbourhoods where people can live, work, and shop in close proximity.

10. **Traffic Management Systems**: Implement smart traffic management systems to reduce congestion and improve the efficiency of road networks.

11. **Green Freight**: Implement strategies to make freight transport more efficient, such as improving aerodynamics, reducing idling, and using more efficient tires.

12. **Sustainable Aviation**: Promote research into and adoption of sustainable aviation technologies, such as more efficient engines, lightweight materials, and sustainable biofuels.

13. **Maritime Shipping Improvements**: Enhance the energy efficiency of ships, optimize routes, and transition to cleaner fuels to reduce the environmental impact of maritime shipping.

14. **Road Pricing or Congestion Charges**: Implement charges for driving in congested areas or during peak times to discourage unnecessary travel and reduce congestion.

15. **Educate and Raise Awareness**: Increase awareness of the environmental impact of transport and educate people on how they can reduce their transport-related emissions.

These solutions involve a range of strategies, from technology and infrastructure development to behaviour change and policy implementation, all of which can help reduce the environmental impact of transportation.

Data transfer and the Internet

Global data transfer and the increasing size of the internet infrastructure have raised environmental, security, and economic concerns. Below are 15 potential solutions to these issues:

1. **Energy-Efficient Data Centres**: Encourage the design and use of energy-efficient data centres. Cooling systems, in particular, represent a large portion of data centre energy usage and can be made more efficient.

2. **Green Power for Data Centres**: Utilize renewable energy sources to power data centres.

3. **Data Optimization**: Implement data optimization techniques to reduce the amount of data being transferred.

4. **Efficient Hardware**: Promote the use of energy-efficient hardware in both servers and end-user devices.

5. **Edge Computing**: Use edge computing to process data closer to its source, reducing the amount of data that needs to be transferred over long distances.

6. **Virtual Meetings and Conferences**: Encourage the use of virtual meetings instead of physical travel for business or academic conferences.

7. **Network Infrastructure Efficiency**: Improve the energy efficiency of network infrastructure such as routers and switches.

8. **Server Virtualization**: Virtualize servers to maximize their usage and reduce the total number of servers needed.

9. **Improved Algorithms**: Develop and use more efficient algorithms that can process data more quickly and/or with less computational power.

10. **Equipment Recycling Programs**: Implement programs to recycle old equipment, reducing the environmental impact of manufacturing new equipment.

11. **Data Storage Optimization**: Use techniques such as deduplication and compression to reduce the amount of storage needed for data.

12. **Telecommuting**: Encourage remote work to reduce the carbon footprint associated with commuting.

13. **Artificial Intelligence and Machine Learning**: Use AI and ML to optimize data storage, network load, and energy usage in data centres.

14. **Internet Caching**: Use caching to store frequently accessed data closer to the user, reducing the amount of data that needs to be transferred over the network.

15. **Regulation and Policy**: Implement policies and regulations that encourage energy efficiency and data optimization in the tech industry.

Each of these solutions can help manage the environmental and economic impact of growing data transfer and internet infrastructure. They involve strategies across hardware and software design, usage practices, and policy implementations.

Exponential technological development

1. **Lifelong Learning and Reskilling Initiatives**: To address the job displacement caused by rapid technological advancement, there is a need to equip workers with new skills through training programs, online courses, and vocational education.

2. **Regulatory Revisions**: Governments need to update laws and regulations to keep pace with technological change, addressing issues like data privacy, cybersecurity, artificial intelligence ethics, and intellectual property rights for digital content.

3. **Digital Inclusion**: To ensure everyone benefits from technological advancements, governments and corporations should work to bridge the digital divide, ensuring internet and technology access for all, regardless of socioeconomic status or geographic location.

4. **Mental Health Programs**: With the rise of digital technology comes increased screen time and the potential for addiction. Implementing programs that promote mental health awareness, digital detox, and balanced use of technology can help address this issue.

5. **Improved Cybersecurity Measures**: With more data being generated and shared, the threat of cyber-attacks is also increasing. Organizations must prioritize robust cybersecurity measures, while individuals should be educated about safe digital practices.

6. **Public Awareness Campaigns**: Governments, educational institutions, and non-profit organizations should conduct campaigns to inform the public about the potential benefits and drawbacks of new technologies, promoting informed decision-making.

7. **Technology Ethics Education**: Introducing ethics courses in technology-related fields can help professionals understand and address the ethical implications of their work, promoting responsible innovation.

8. **Responsible AI Development**: AI developers should consider the ethical implications of their work, such as algorithmic bias, and strive for transparency and fairness in their AI systems.

9. **Data Privacy Protections**: Strengthening data protection laws and norms can help protect individuals' privacy in an increasingly digital world. Individuals also need to be educated about their rights and how to protect their own data.

10. **E-Waste Management**: Rapid tech advancement leads to more electronic waste. Strategies must be implemented for better e-waste recycling and disposal.

11. **Sustainable Technology Practices**: Technology companies should be encouraged to adopt more sustainable practices, such as reducing energy usage in

data centres and designing products for durability and recyclability.

12. **Universal Basic Income (UBI)**: To counteract job loss due to automation, some propose the implementation of a universal basic income that guarantees a living wage for all citizens, regardless of employment.

13. **Tech Worker Rights Advocacy**: Ensuring fair wages, reasonable hours, and the right to unionize for tech workers, including gig workers, can help to alleviate labour issues in the tech industry.

14. **Public-Private Partnerships (PPPs)**: Governments and private organizations can partner to develop, regulate, and distribute technology in a manner that maximizes societal benefits and minimizes drawbacks.

15. **Mental Health Tech Regulation**: Technologies such as social media have significant impact on mental health. Stricter regulation on these platforms, including controls on content and usage habits, could mitigate potential negative impacts.

Soil degradation

Soil regeneration is crucial for maintaining productive agricultural systems and for mitigating climate change. Here are 15 potential strategies for regenerating global soils:

1. **Cover Crops**: Planting of cover crops to protect the soil from erosion and to add organic matter.

2. **Crop Rotation**: Rotating crops to improve soil fertility and break the cycle of pests and diseases.

3. **Composting**: Adding compost to soils to increase organic matter and boost soil fertility.

4. **Reduced Tillage/No-Till Farming**: Reducing soil disruption preserves soil structure, encourages microbial activity, and minimizes erosion.

5. **Agroforestry**: Incorporating trees into farming systems can improve soil fertility, prevent erosion, and sequester carbon.

6. **Permaculture**: Designing agricultural systems to mimic natural ecosystems can increase soil health and resilience.

7. **Mulching**: Applying a layer of organic material on the soil surface can conserve moisture, reduce erosion, and improve soil health.

8. **Managed Grazing**: Implementing grazing strategies that mimic natural, wild herd behaviours can improve grass growth and soil health.

9. **Biodynamic Farming**: This holistic approach to agriculture emphasizes the interrelationships of the soil, plants, and animals as a self-sustaining system.

10. **Organic Farming**: Avoiding synthetic fertilizers and pesticides can enhance soil biodiversity and long-term fertility.

11. **Adding Biochar**: Incorporating biochar, a form of charcoal, into the soil can sequester carbon and improve soil fertility.

12. **Polyculture and Biodiversity**: Cultivating a diversity of crops can improve soil health and resilience.

13. **Terracing**: On slopes, terracing can reduce erosion and enhance water retention.

14. **Conservation Buffers**: Planting strips of vegetation between crops can reduce runoff and erosion.

15. **Green Manure**: Growing plants to be ploughed back into the soil adds organic matter and nutrients, improving soil health.

Each of these strategies can contribute to soil regeneration. However, the most effective approach will depend on the specific context, including the local climate, soil type, and agricultural practices.

Solutions – UN 17 SDGs

Inequality

1. **Progressive Taxation**: The rich are taxed more than the poor, reducing income inequality. Capital gains, income from investments, should also be taxed more heavily. This policy attempts to redistribute wealth from the wealthy to the less privileged.

2. **Universal Basic Income (UBI)**: This policy ensures a basic level of income for all, regardless of employment status. UBI is considered a tool to alleviate poverty and provide financial stability, reducing wealth disparity.

3. **Education Investment**: By ensuring everyone has access to quality education, individuals can gain the skills needed to secure well-paying jobs. This can reduce wealth inequality by providing equal opportunities.

4. **Minimum Wage Increase**: Raising the minimum wage can help bridge the income gap between the highest and lowest earners.

5. **Inheritance Tax**: High inheritance taxes can help prevent the accumulation of wealth in certain families and promote a more equal distribution of wealth.

6. **Wealth Tax**: Taxing a person's net worth above a certain threshold can help reduce massive wealth concentration and create a more equal society.

7. **Access to Quality Healthcare**: Wealth inequality often leads to health inequality. By ensuring equal access to healthcare, we can reduce the disparities that wealth inequality generates.

8. **Labor Market Reforms**: Policies that promote worker's rights, like strong labour unions and protections against unfair dismissals, can help to distribute wealth more evenly.

9. **Affordable Housing Initiatives**: Implementing policies that increase the availability of affordable housing can help lessen the wealth gap, as housing is a significant factor of wealth.

10. **Financial Literacy and Access**: Increasing access to financial services and improving financial literacy can help individuals better manage their personal finances, reduce debt, and increase savings, thereby reducing wealth inequality.

1. **Public Awareness Campaigns**: Raise public awareness about wealth inequality and its impact on society. This could help build public pressure on political leaders to address the issue.

2. **Voter Education**: Educate voters about wealth inequality and the policy options to address it. An informed electorate can vote for leaders and policies that prioritize reducing inequality.

3. **Grassroots Activism**: Encourage and support grassroots movements aimed at reducing wealth inequality. These movements can help shift public opinion and put pressure on political leaders.

4. **Public-Private Partnerships**: Encourage collaborations between governments, private sector companies, and non-profit organizations to implement initiatives aimed at reducing wealth inequality.

5. **Encourage Political Participation**: Mobilize marginalized and disenfranchised groups to participate in the political process. This could shift the balance of power and lead to more policies aimed at reducing wealth inequality.

6. **Policy Advocacy**: Advocate for specific policies that reduce wealth inequality, such as progressive taxation, higher minimum wages, and increased public investment in education and healthcare.

7. **Promote Social Responsibility Among Corporations**: Encourage corporations to pay fair wages, offer good benefits, and contribute to their communities. This could help reduce wealth inequality even in the absence of government action.

8. **Transparency and Accountability**: Advocate for greater transparency and accountability in government, which can help expose corruption and make it more likely that public resources are used to address wealth inequality.

9. **International Cooperation**: Work with international organizations to address wealth inequality. This can put additional pressure on governments to act and provide them with practical tools and policies.

10. **Political Campaign Finance Reform**: Advocate for reforms to political campaign financing to reduce the influence of wealthy donors and interest groups. This

could make it more likely that political leaders will act in the interests of all their constituents, not just the wealthy ones.

These solutions all aim to shift the balance of power and create conditions in which political leaders are more likely to take action to reduce wealth inequality. However, they require sustained effort and broad-based support to be successful.

Food

1. **Sustainable Agriculture**: Promote farming practices that protect and improve soil health, such as cover cropping, crop rotation, and reduced tillage, to maintain soil fertility and productivity.

2. **Agroforestry**: Implement agroforestry systems, which combine trees with crops and/or livestock, to enhance soil fertility, increase biodiversity, and sequester carbon.

3. **Precision Agriculture**: Use precision farming techniques and technologies to apply water, fertilizer, and other inputs more accurately and efficiently, reducing waste and minimizing environmental impact.

4. **Biofertilizers**: Promote the use of biofertilizers, such as nitrogen-fixing bacteria and compost, to replace or supplement traditional chemical fertilizers.

5. **Crop Diversity**: Encourage the cultivation of a diverse range of crops, including native and climate-resilient varieties, to enhance agricultural resilience and soil health.

6. **Efficient Irrigation**: Promote the use of efficient irrigation techniques, such as drip irrigation, to save water and prevent soil erosion and nutrient leaching.

7. **Regenerative Agriculture**: Implement regenerative agriculture practices, which restore and enhance soil health, sequester carbon, and improve resilience to climate change.

8. **Food Waste Reduction**: Reduce food waste at all stages of the food supply chain to increase the overall efficiency of food production and distribution.

9. **Education and Training**: Provide education and training to farmers on sustainable farming practices, climate change adaptation, and soil conservation.

10. **Research and Development**: Invest in research and development of new crops, farming practices, and technologies that can withstand or mitigate the impacts of climate change, soil degradation, and dwindling fertilizer supplies.

Each of these solutions offers a way to address the challenges of providing adequate food in the face of soil degradation, lack of fertilizer feedstocks, and climate change. Implementing them will require concerted efforts at all levels, from individual farmers to international organizations, and a strong commitment to sustainable and resilient food systems.

Personal health

The decline in global personal health and fitness is a complex issue influenced by various factors, including lifestyle, diet, mental health, environment, and more. Here are 15 solutions that can help address this issue:

1. **Promote Regular Exercise**: Encourage people to incorporate regular physical activity into their lives. This could involve developing programs that encourage walking, cycling, and other types of exercise.

2. **Healthy Eating Campaigns**: Develop public health campaigns that promote the benefits of a balanced diet, rich in fruits, vegetables, lean proteins, and whole grains.

3. **Mental Health Awareness**: Raise awareness about the importance of mental health and provide resources to help people manage stress, anxiety, and other mental health issues.

4. **Accessible Healthcare**: Improve access to quality healthcare for all individuals, regardless of their economic status.

5. **Workplace Wellness Programs**: Encourage businesses to implement wellness programs that provide employees with resources for maintaining their physical and mental health.

6. **Community Sports and Fitness Programs**: Create and promote affordable and accessible community sports programs and fitness classes.

7. **Education**: Incorporate health and nutrition education into school curriculums from an early age.

8. **Active Transportation**: Encourage walking, cycling, and other forms of active transportation through urban planning and infrastructure development.

9. **Food Policies**: Implement policies that promote the availability and affordability of healthy food options, and restrict marketing of junk food, especially to children.

10. **Preventive Health Screenings**: Encourage regular preventive health screenings to catch potential health problems early.

11. **Tobacco and Alcohol Control**: Implement policies to reduce the consumption of tobacco and alcohol, such as higher taxes, restrictions on advertising, and public health campaigns.

12. **Green Spaces**: Create and maintain parks and other green spaces where people can exercise and relax.

13. **Public-Private Partnerships**: Encourage partnerships between governments, NGOs, and private companies to promote health and fitness.

14. **Telemedicine**: Leverage technology to provide remote healthcare services, especially in underserved areas.

15. **Incentive Programs**: Create programs that offer incentives for individuals to maintain their health, such as reduced insurance premiums for regular exercise or health checks.

Remember that improving global personal health and fitness requires a multifaceted approach that addresses physical, mental, and social aspects of health. It also requires collaboration across various sectors, including healthcare, education, transportation, and urban planning.

Disease

list the top 15 solutions to problems generated by poor global health

1. **Universal Healthcare**: Ensuring access to affordable and quality healthcare for all, irrespective of socio-economic status.

2. **Preventive Care**: Investing in preventive care, including vaccinations, screenings, and public health awareness campaigns.

3. **Clean Water and Sanitation**: Guaranteeing access to clean water and proper sanitation facilities, which are crucial for preventing many diseases.

4. **Nutrition**: Promoting access to healthy food and nutrition education to prevent diet-related diseases like obesity and diabetes.

5. **Health Education**: Educating the public about health and hygiene practices, disease prevention, and the importance of regular medical check-ups.

6. **Research and Development**: Investing in research for the development of new treatments, therapies, and vaccines.

7. **Mental Health Services**: Providing resources and services for mental health care, often a neglected aspect of overall health.

8. **Addressing Social Determinants of Health**: Working on factors like poverty, education, and housing, which have significant impacts on health.

9. **Physical Activity**: Promoting regular physical activity as a key part of maintaining good health.

10. **Reducing Health Disparities**: Addressing disparities in health access and outcomes based on race, ethnicity, gender, socio-economic status, and geographical location.

11. **Tobacco and Alcohol Control**: Implementing measures to control tobacco and alcohol use, major contributors to global disease burden.

12. **Healthcare Workforce**: Training and retaining a strong healthcare workforce, particularly in regions that lack adequate healthcare services.

13. **Digital Health**: Leveraging technology for healthcare delivery, especially for remote or underserved populations.

14. **Emergency Preparedness**: Building robust systems for responding to health emergencies like epidemics and pandemics.

15. **International Cooperation**: Collaborating globally for knowledge exchange, resource sharing, and coordinated responses to global health challenges.

These solutions require significant investment and collaboration among government entities, non-governmental organizations, healthcare providers, and the public. However, they can greatly enhance global health and reduce the burden of disease.

Education

1. **Increased Accountability**: A well-educated population is likely to be more aware of their rights and demand accountability from their leaders, increasing political pressure.

2. **Critical Thinking**: Education promotes critical thinking, which can lead to questioning established systems, norms, and policies, potentially challenging the status quo.

3. **Informed Voting**: A well-educated populace can make informed voting decisions, which may not always align with the interests of the political elite.

4. **Policy Scrutiny**: Education enables citizens to better understand and critique policy proposals, limiting the ability of politicians to manipulate public opinion.

5. **Social Mobility**: Education is a key driver of social mobility, which can threaten established hierarchies and power structures.

6. **Demand for Transparency**: Educated citizens may demand greater transparency and openness in government, which can expose corruption or misconduct among the political elite.

7. **Higher Expectations**: A well-educated population may have higher expectations of their political leaders, placing additional demands on them.

8. **Community Mobilization**: Education can empower citizens to organize and mobilize for political causes, challenging the control of the political elite.

9. **Reduced Dependency**: Education can reduce economic dependency on the state, allowing citizens to be more independent and possibly less influenced by political rhetoric.

10. **Promotion of Equality**: Education promotes social equality and inclusivity, which can challenge systems of privilege that often benefit the political elite.

11. **Political Activism**: An educated populace is more likely to engage in political activism, advocating for changes that may not align with the interests of the current political elite.

12. **Demand for Justice**: Educated individuals are typically more aware of social and economic injustices and may demand corrective measures, potentially challenging those who benefit from existing inequities.

13. **Understanding of Rights**: An educated population is more likely to understand and assert their rights, which can lead to resistance against authoritarian or unfair practices.

14. **Global Awareness**: Education often enhances understanding of global issues and contexts, making it harder for leaders to promote insular or nationalist ideologies.

15. **Increased Political Participation**: A more educated population may be more likely to participate in politics at all levels, not just voting, which can result in a shift in power dynamics.

16. **Empowerment of Marginalized Groups**: Education can empower traditionally marginalized groups, altering societal structures and potentially threatening established hierarchies.

17. **Data Literacy**: A well-educated populace is more likely to be data literate and better equipped to understand and interpret the use of statistics in public discourse, making it harder for political leaders to manipulate or misrepresent information.

18. **Support for Institutional Checks and Balances**: With a better understanding of how government works, an educated public is more likely to support systems that limit abuses of power.

19. **Promotion of Civil Liberties**: Education increases awareness of civil liberties, potentially leading to increased demand for freedom of speech, press, and assembly.

20. **Shifts in Social Values**: As education levels rise, social values can shift, including towards greater acceptance of diversity, equality, and democratic norms, which can challenge leaders who rely on division or demagoguery.

Again, it's important to note that these factors, while potentially challenging to political elites, contribute to a healthier, more responsive political environment and are crucial for the long-term stability and prosperity of any society.

Peace, justice and strong institutions

Creating world peace, fostering justice, and building strong, honest institutions is a complex process that requires a comprehensive, multidimensional approach. Here are 15 potential solutions that could contribute to these goals:

1. **Education**: Promote access to quality education for all. Education broadens perspectives, breaks down barriers, and fosters understanding between different cultures and social groups.

2. **Dialogue and Diplomacy**: Prioritize dialogue and diplomacy as the primary means of resolving conflicts between nations, groups, or individuals.

3. **Inclusive Governance**: Create governance structures that are inclusive, ensuring that all voices are heard, not just those of the elite or majority.

4. **Transparency**: Encourage transparency in institutions to build public trust and to prevent corruption.

5. **Rule of Law**: Uphold the rule of law to ensure fair treatment for all citizens, regardless of their social, economic, or political status.

6. **Human Rights**: Promote and protect human rights at all levels, and hold those who violate these rights accountable.

7. **Conflict Resolution Education**: Incorporate conflict resolution and peace education into school curriculums and adult education programs.

8. **Economic Equality**: Work towards economic equality by implementing policies that reduce extreme wealth disparities and promote economic opportunities for all.

9. **Community Engagement**: Foster strong, supportive communities where people are actively involved in decision-making processes.

10. **Independent Media**: Support an independent and diverse media sector that can hold power to account and provide citizens with accurate, balanced information.

11. **Access to Justice**: Ensure that all individuals have access to justice, including legal representation and fair trials.

12. **International Cooperation**: Strengthen international institutions and encourage cooperation between countries to address global challenges.

13. **Respect for Diversity**: Promote respect for cultural, religious, and social diversity to foster mutual understanding and peaceful coexistence.

14. **Decentralization of Power**: Avoid the concentration of power by decentralizing governmental structures, giving more power to local and regional bodies.

15. **Sustainable Development**: Promote sustainable development to ensure that our actions do not harm future generations, create conflict over resources, or exacerbate social inequalities.

Each of these solutions requires the active participation of individuals, communities, governments, and international organizations. While progress may be slow and there will undoubtedly be challenges along the way, these steps could contribute significantly to achieving world peace, justice, and strong, honest institutions.

Globalisation

Globalization has brought a range of challenges for the world's working population, including job displacement due to automation and outsourcing, income inequality, and increased competition. Here are 15 potential solutions to these problems:

1. **Education and Lifelong Learning**: Enhance education systems and promote lifelong learning to help workers adapt to changing job markets.

2. **Skills Training and Retraining**: Implement programs that help workers develop the skills needed for the jobs of the future.

3. **Digital Literacy**: Promote digital literacy to ensure workers can compete in increasingly digital economies.

4. **Social Safety Nets**: Strengthen social safety nets to protect workers who lose their jobs or face economic hardship.

5. **Labor Rights Protection**: Strengthen laws and regulations to protect workers' rights in all countries.

6. **Income Redistribution Policies**: Implement policies that promote income redistribution, such as progressive taxation and increased minimum wages.

7. **Local Economic Development**: Support local businesses and industries to help create jobs and stimulate local economies.

8. **Telecommuting Opportunities**: Encourage companies to offer remote work opportunities, enabling people to work from anywhere.

9. **Diversity and Inclusion Initiatives**: Promote diversity and inclusion in the workplace to ensure all individuals have equal opportunities.

10. **Sustainable Development Goals (SDGs)**: Encourage companies and governments to align their policies and practices with the United Nations' SDGs.

11. **Public-Private Partnerships**: Promote partnerships between governments, NGOs, and the private sector to address labour market challenges.

12. **Regulation of Global Corporations**: Implement policies and regulations to prevent tax avoidance, exploitation, and other harmful practices by multinational corporations.

13. **Fair Trade Policies**: Support fair trade to ensure that workers in all parts of the supply chain receive fair wages and work in good conditions.

14. **Healthcare and Retirement Benefits**: Establish systems to provide healthcare and retirement benefits to all workers, including those in informal or gig economy jobs.

15. **Community Development Programs**: Invest in community development programs to create jobs and stimulate economic growth at the local level.

These solutions require a collaborative and concerted effort from governments, companies, NGOs, and workers themselves. With the right policies and support systems in place, the challenges of globalization can be effectively addressed.

Solutions – Societal Behaviour

Over-consumption of energy and resources

1. **Sustainable Production**: Encourage companies to adopt sustainable production methods, reducing the overall environmental impact of goods and services.

2. **Consumer Education**: Educate consumers about the environmental impact of overconsumption and the benefits of sustainable consumption.

3. **Repair and Reuse**: Promote a culture of repairing, reusing, and upcycling items instead of discarding them after a single use. This can reduce the demand for new products and thus, overconsumption.

4. **Eco-Labelling**: Implement product labelling that clearly communicates the environmental impact of goods, enabling consumers to make more sustainable choices.

5. **Green Taxes**: Implement tax policies that incentivize sustainable consumption and production. For instance, higher taxes on plastic products can encourage the use of alternatives.

6. **Regulating Advertising**: Implement strict regulations on advertising to reduce the influence of consumerism and promote responsible consumption.

7. **Sustainable Public Procurement**: Governments can lead by example by prioritizing the procurement of goods and services that are sustainable, thereby driving demand for such products.

8. **Zero-Waste Initiatives**: Implement policies aimed at achieving zero waste through recycling and composting programs, reducing the need for new materials.

9. **Sustainable City Planning**: Develop cities in a way that encourages sustainable living, for example, by creating spaces for community gardens and providing infrastructure for cycling and public transportation.

10. **Circular Economy**: Encourage a shift towards a circular economy where resources are reused and recycled rather than being discarded, reducing the demand for new resources and thus, overconsumption.

Each of these solutions aims to reduce the environmental impact of overconsumption, but implementing them requires commitment and coordination from governments, businesses, and individuals.

1. **Efficient Use and Conservation**: Develop and implement strategies for more efficient use of mineral resources, including energy-efficient technologies and reduced waste in production processes.

2. **Recycling and Reuse**: Increase recycling and reuse of mineral resources in products to reduce the need for new mining and extraction.

3. **Substitution**: Research and develop substitute materials that can perform the same functions as certain critical minerals.

4. **Diversification of Supply**: Diversify supply chains and reduce dependence on a few countries for critical minerals.

5. **Geological Surveys and Exploration**: Invest in geological surveys and exploration to discover new deposits of critical minerals.

6. **Sustainable Mining Practices**: Adopt and enforce sustainable mining practices to minimize environmental impact and ensure the longevity of mining operations.

7. **Enhanced International Cooperation**: Foster international cooperation for the responsible and sustainable use of global mineral resources.

8. **Investment in Research and Development**: Increase investment in R&D for new technologies that require fewer or different materials.

9. **Strategic Reserves**: Establish strategic reserves of critical minerals to buffer against supply disruptions.

10. **Transparent and Fair Trade Policies**: Encourage transparent and fair trade policies that ensure equitable access to mineral resources.

11. **Legal Framework for Deep-Sea Mining**: Develop a comprehensive legal framework for deep-sea mining that balances resource extraction with environmental protection.

12. **Urban Mining**: Implement urban mining by extracting metals from discarded products and waste.

13. **Education and Training**: Educate and train the workforce in efficient and sustainable use of mineral resources.

14. **Advanced Extraction Techniques**: Develop and use advanced extraction techniques that can recover more resources from ore.

15. **Promote Local Production and Processing**: Encourage local production and processing of critical minerals to reduce dependency on foreign resources.

The challenge lies in balancing the need for these critical minerals with the environmental, social, and economic impacts of their extraction and use. All these strategies can play a role in ensuring a steady supply of critical minerals in the future while minimizing negative impacts.

Misinformation, censorship and mass media control

Dealing with censorship, media manipulation, and misinformation is a complex challenge that requires comprehensive solutions on multiple levels. Here are some potential strategies:

1. **Media Literacy Education**: Enhance the public's ability to critically evaluate media content. This involves understanding how media works, how to discern reliable sources, and how to recognize manipulative techniques.

2. **Fact-Checking Organizations**: Support fact-checking organizations that can help verify information and expose false or misleading narratives. These can be independent or connected to media outlets.

3. **Transparency in Algorithms**: Demand transparency from social media companies about their content promotion algorithms. If users understand how content is promoted, they may be better equipped to spot manipulated narratives.

4. **Regulation of Social Media Platforms**: Regulate social media platforms to ensure they take responsibility for the spread of misinformation and manipulation on their platforms.

5. **Promoting Quality Journalism**: Support traditional journalism, where there is an emphasis on accuracy, fairness, and transparency. This could be through direct subscription, donations, or policy measures.

6. **Platform Accountability**: Hold digital platforms accountable for the content they host, with penalties for spreading dangerous or harmful misinformation.

7. **Open Access to Information**: Counteract censorship by promoting open access to information, using platforms that are difficult to censor and technologies that protect privacy and security, like VPNs or blockchain.

8. **Public Interest Technology**: Develop technology solutions that prioritize public interest and protect democratic discourse. This could include tools to trace the source of information or to track the spread of a piece of news.

9. **International Cooperation**: Work across borders to tackle the global challenge of misinformation and media manipulation, with shared standards and coordinated actions.

10. **Digital Citizen Initiatives**: Encourage citizen initiatives that promote responsible digital behaviour, like sharing verified information only and reporting misinformation.

11. **Whistleblower Protections**: Strengthen protections for whistleblowers who expose manipulation, censorship, or misinformation within media organizations or tech companies.

12. **Research**: Invest in research to understand the effects of misinformation and media manipulation, as well as to develop effective strategies to combat them.

13. **Promote Independent Media**: Foster a diverse and independent media landscape to provide a range of perspectives and reduce the power of any single media entity.

14. **User Controls over Personal Data**: Enable users to control how their personal data is used to target content, including news and advertisements.

15. **Civic Education**: Include elements of digital citizenship, like recognizing misinformation and understanding the role of free press, in civic education.

Remember, addressing these issues requires a combination of many strategies, along with the political will, resources, and public support to implement them effectively.

Business as usual and the military industrial complex

The global "business as usual" approach and the influence of the military-industrial complex can contribute to a variety of social, economic, and environmental issues. Here are some strategies that could help to address these challenges:

1. **Sustainable Development**: Businesses should shift toward sustainable development models that prioritize environmental and social well-being alongside economic growth.

2. **Regulation of Lobbying**: Tighten regulations on lobbying to reduce the undue influence of large corporations, including defence contractors, on government policy.

3. **Transparency**: Increase transparency in military budgets and procurement contracts to prevent corruption and overspending.

4. **Corporate Social Responsibility**: Encourage companies to prioritize corporate social responsibility, which may include environmental stewardship, fair labour practices, and community development.

5. **Circular Economy**: Encourage the adoption of circular economy principles, which aim to minimize waste and make the most of resources.

6. **Diversification of Economy**: Reduce dependence on the military-industrial complex by diversifying the economy and investing in other sectors.

7. **Economic Conversion**: Support economic conversion, the process of transitioning from military to civilian industries, which can create jobs and stimulate sustainable development.

8. **Peace and Diplomacy**: Prioritize diplomacy and peaceful conflict resolution over military intervention, reducing the need for a large military-industrial complex.

9. **Incentives for Green Practices**: Implement fiscal incentives for businesses to adopt environmentally friendly practices.

10. **Education and Research**: Invest in education and research to innovate new, sustainable technologies and practices.

11. **Reform Intellectual Property Rights**: Reforms in intellectual property rights can stimulate innovation and accessibility of sustainable technologies.

12. **Global Cooperation**: Encourage international cooperation and treaties that work towards mutual disarmament and climate action.

13. **Democratic Control**: Increase democratic control and oversight over military decisions to ensure they align with public interest.

14. **Public-Private Partnerships**: Encourage partnerships between governments, private companies, and non-profit organizations to address societal issues.

15. **Invest in Renewable Energy**: Redirect some of the resources currently directed towards the military-industrial complex towards the development and

deployment of renewable energy solutions, which can also contribute to energy security.

The application of these strategies would require careful planning, strong political will, and international cooperation, but they can go a long way in creating a more sustainable, equitable, and peaceful world.

Excessive bureaucracy

1. **Process Simplification**: Streamline processes to remove redundant steps and make the system more efficient.

2. **Digitization**: Implement digital solutions to replace manual, paper-based processes. This can reduce errors, speed up processes, and make services more accessible.

3. **Transparent Governance**: Foster transparency to hold bureaucracies accountable, reduce corruption, and encourage efficiency.

4. **Performance Metrics**: Implement performance metrics to evaluate the efficiency of different departments and reward high-performing units or individuals.

5. **Decentralization**: Delegate authority to lower levels of the organization to speed up decision-making and reduce bottlenecks.

6. **Training**: Provide regular training to staff to improve their efficiency, understand their roles better, and reduce redundancy in work processes.

7. **Automation**: Use technology to automate routine tasks, which can save time, reduce errors, and free up staff to focus on more complex tasks.

8. **Customer-Oriented Services**: Refocus the bureaucracy to be more service-oriented, emphasizing the needs of the citizens or customers. This could involve creating user-friendly interfaces, reducing wait times, and improving overall customer service.

9. **Regulatory Review**: Periodically review regulations to remove outdated or unnecessary ones and reduce red tape.

10. **Participatory Decision-Making**: Involve citizens or end-users in the decision-making process. This can improve the relevance and acceptance of decisions and reduce the need for heavy bureaucracy.

These solutions all aim to reduce the inefficiencies and frustrations often associated with bureaucracy. However, they require strong leadership, a commitment to change, and the necessary resources for implementation.

1. **Regulatory Oversight**: Establish more robust regulation and oversight of certification processes to ensure their credibility, effectiveness, and consistency.

2. **Transparency**: Promote transparency in certification standards and methods to enhance public trust and encourage responsible behaviour by businesses.

3. **Accurate Carbon Accounting**: Improve the accuracy of carbon accounting methods to ensure that offsets and net zero claims accurately reflect real-world emissions reductions.

4. **Addressing Scope 3 Emissions**: Encourage businesses to account for and address Scope 3 emissions (those that occur in their supply chains) in their ESG, carbon offsetting, and net zero efforts.

5. **Investor Education**: Educate investors about the nuances and potential pitfalls of ESG, carbon offsetting, and net zero certifications, helping them make more informed decisions.

6. **Promote Actual Emission Reductions**: Encourage companies to prioritize actual emissions reductions within their operations, rather than relying primarily on offsets or net zero claims.

7. **Third-Party Verification**: Encourage or require third-party verification of certifications to ensure that they are reliable and that companies are held accountable.

8. **Standardization**: Work towards standardized, globally recognized certification schemes to prevent "greenwashing" and make comparisons easier for consumers and investors.

9. **Include Social and Governance Factors**: Ensure that the social and governance aspects of ESG certifications are not neglected in favour of environmental factors, helping to create a more balanced and comprehensive view of sustainability.

10. **Public-Private Partnerships**: Establish partnerships between public and private sector organizations to promote best practices, improve certification processes, and provide the necessary resources for businesses to achieve their ESG and carbon neutrality goals.

While ESG, carbon offsetting, and net zero certifications can play a vital role in the transition to a more sustainable economy, they need to be implemented in a robust, transparent, and accountable way to avoid greenwashing and ensure real progress is being made.

Social cohesion

1. **Education**: Provide education programs focusing on diversity, equality, and human rights. Teach the historical, social, and cultural context of different groups to foster understanding and respect.

2. **Community Outreach Programs**: Encourage community interactions and events to help different groups learn about each other and break down barriers.

3. **Legislation and Policies**: Enforce laws and policies that protect minority groups and penalize discriminatory behaviour. This could include stricter hate crime legislation.

4. **Representation**: Promote diversity in all levels of leadership, media, and education to ensure that various cultural, racial, and ethnic groups are represented.

5. **Public Campaigns**: Launch public campaigns that promote the positive aspects of a multicultural society and highlight the harmful effects of xenophobia.

6. **Dialogue and Mediation**: Encourage dialogue and conflict resolution initiatives to help resolve misunderstandings and disputes between different cultural, ethnic, or racial groups.

7. **Anti-discrimination Training**: Implement anti-discrimination training in workplaces, schools, and community centres to equip individuals with the skills to counteract bias and prejudice.

8. **Support Services for Victims**: Provide support services, such as counselling, legal aid, and hotlines, for victims of xenophobia and discrimination.

9. **Promote Inclusive Policies**: Governments and organizations should encourage policies that facilitate integration, such as language training, equal access to services, and fair employment practices.

10. **International Cooperation**: Collaborate with international organizations and countries to share best practices and establish global norms and policies against xenophobia and social division.

Each of these solutions aims to foster an inclusive society where all people are treated equally and with respect. However, implementing these solutions requires consistent commitment from individuals, communities, organizations, and governments.

Global debt

Excessive world debt, both public and private, can lead to a range of economic problems, including financial crises, reduced public spending capacity, and potential insolvency. Here are 15 potential solutions to these challenges:

1. **Fiscal Responsibility**: Encourage governments to exercise fiscal responsibility, avoiding excessive borrowing and managing public resources effectively.

2. **Debt Restructuring**: Restructure existing debts, lengthening maturity periods or reducing interest rates to make repayments more manageable.

3. **Economic Growth Policies**: Implement policies that foster sustainable economic growth, increasing the capacity to repay debts.

4. **Debt Forgiveness**: In some extreme cases, creditors (often international institutions or other countries) may need to forgive some of a debtor country's obligations.

5. **Regulation of Financial Markets**: Strengthen regulation of financial markets to avoid risky lending and borrowing practices.

6. **Transparent Borrowing Practices**: Ensure that borrowing practices are transparent, avoiding hidden debts or obligations.

7. **Austerity Measures**: Implement spending cuts or tax increases to reduce public debt, though this must be done carefully to avoid harming economic growth or disproportionately affecting the vulnerable.

8. **Monetary Policy**: Use monetary policy tools, such as interest rate adjustments, to manage debt levels.

9. **Strengthen Tax Collection**: Improve tax collection systems to increase government revenue and reduce the need for borrowing.

10. **Invest in Education and Infrastructure**: Invest in human and physical capital to improve productivity and economic growth potential.

11. **International Cooperation**: Strengthen international cooperation to manage global economic challenges, including excessive debt.

12. **Early Warning Systems**: Develop early warning systems to detect unsustainable debt levels and respond proactively.

13. **Debt Swap**: Consider debt-for-nature or debt-for-development swaps, where part of a country's foreign debt is forgiven in exchange for investments in environmental conservation or development projects.

14. **Debt Management Office**: Establish a dedicated debt management office to oversee borrowing and repayment strategies.

15. **Public Debt Audits**: Conduct regular audits of public debts to ensure transparency and accountability in how borrowed funds are used.

These solutions involve a range of strategies, from policy changes to institutional reforms. While not all of these may be applicable in every context, they provide a variety of approaches to managing and reducing excessive world debt.

Global financial system

list the top 10 solutions to problems generated by
manipulation of financial markets

1. **Transparency and Disclosure Requirements**:
   Improved transparency and comprehensive disclosure
   of financial information can make it more difficult for
   manipulative practices to occur unnoticed.

2. **Robust Regulatory Oversight**: Regulatory agencies
   need the resources and power to monitor and police the
   market effectively. This can include real-time
   surveillance of trading activity and the ability to swiftly
   penalize those who engage in manipulative practices.

3. **Stronger Penalties for Market Manipulation**: Higher
   fines, penalties, and the possibility of jail time can serve
   as a deterrent against market manipulation.

4. **Clearer Regulations**: Specific, clear, and
   comprehensive rules against market manipulation help
   create a fairer marketplace.

5. **International Cooperation**: Given the global nature of
   financial markets, countries need to work together to
   combat market manipulation. This can include sharing
   information and harmonizing regulations.

6. **Improved Market Infrastructure**: Better market
   infrastructure, such as upgraded trading systems, can
   help prevent market manipulation.

7. **Investor Education**: By teaching investors how to spot
   signs of market manipulation, they can be empowered
   to make more informed decisions, report suspicious

activity, and avoid falling victim to manipulative practices.

8. **Better Auditing Practices**: Regular, independent audits can help detect and deter manipulation.

9. **Whistleblower Incentives**: Incentives for whistleblowers, such as financial rewards and protections against retaliation, can encourage more people to report market manipulation.

10. **Algorithmic Accountability**: In the age of high-frequency trading, it's important to have measures in place to ensure algorithms are not being used to manipulate markets. This can involve regular audits of algorithms and stricter rules for algorithmic trading.

These solutions are designed to build trust in the marketplace, protect investors, and promote economic stability. However, implementation will depend on the political will, available resources, and specific circumstances of each jurisdiction.

1. **Fiscal Responsibility**: Governments need to practice fiscal discipline, prioritize spending, and create balanced budgets to control debt levels and reduce the need for currency debasement.

2. **Debt Restructuring**: In some situations, it may be necessary for indebted countries to restructure their debts, extending repayment periods, reducing interest rates, or even negotiating debt forgiveness.

3. **Increase Tax Revenue**: Implementing progressive tax systems and combating tax evasion and avoidance can help increase government revenue, reducing the need for excessive borrowing.

4. **Monetary Policy Reform**: Central banks should implement responsible monetary policies to prevent inflation and maintain the value of the currency.

5. **Encourage Economic Growth**: Policies that stimulate economic growth can increase national income and thereby the ability to repay debts.

6. **Enhanced Transparency**: Countries should enhance transparency in public finance management to ensure borrowed funds are used efficiently and for productive purposes.

7. **Financial Education**: Governments should promote financial literacy among their citizens to foster a culture of saving and investing rather than excessive borrowing.

8. **Promote Fair Trade**: Encourage and engage in fair trade policies to boost export income and improve the balance of trade, reducing the need for debt.

9. **International Cooperation**: International financial institutions and creditor nations should work together with debtor nations to provide support and develop strategies for managing and reducing debt.

10. **Promote Sustainable Development**: Focus on sustainable and inclusive economic development strategies that take into account environmental, social, and economic factors, reducing the need for debt-driven growth.

Excessive global debt and currency debasement can have serious consequences, including economic instability, reduced investor confidence, and potential crises. Addressing these challenges requires a balanced and coordinated approach at both the national and international level.

Giant Tech Monopolies and FANG

FANG is an acronym that stands for Facebook, Amazon, Netflix, and Google, some of the most influential tech companies. Their dominance has raised issues around monopoly power, privacy, data security, and misinformation. Here are 15 potential solutions to these problems:

1. **Antitrust Laws**: Strengthen and enforce antitrust laws to prevent monopolistic practices and promote competition.

2. **Data Portability**: Promote data portability, allowing users to move their data from one platform to another more easily.

3. **Privacy Regulations**: Implement strong privacy regulations to protect user data and ensure informed consent.

4. **Transparency**: Require these companies to be more transparent about their algorithms, data collection practices, and advertising operations.

5. **Cybersecurity Standards**: Enforce robust cybersecurity standards to protect user data and prevent breaches.

6. **Tax Reform**: Ensure these companies pay their fair share of taxes where their products and services are consumed.

7. **Public Oversight**: Establish public oversight bodies to monitor the activities and practices of these tech giants.

8. **Digital Literacy**: Promote digital literacy so that users understand how their data is used and how to navigate digital spaces safely.

9. **Fact-Checking and Misinformation Control**: Collaborate with independent fact-checkers and implement strict policies against spreading misinformation.

10. **Promoting Competitors**: Encourage the growth of competitors in the tech industry, for instance, through grants or favourable regulation.

11. **Breaking Up Companies**: Consider breaking up these companies if they are found to be stifling competition and causing harm to consumers.

12. **Neutral Platform Regulation**: Implement regulations to ensure that these platforms act as neutral spaces and do not unfairly favour their own products or services.

13. **Open Source**: Encourage open-source alternatives to proprietary software and platforms.

14. **User Control Over Data**: Enable users to have more control over their data, including the ability to view, move, or delete it.

15. **Promoting Ethical Practices**: Encourage and reward ethical business practices within these companies.

These solutions require cooperation from various stakeholders, including governments, the tech industry, civil society, and users. They aim to create a more equitable digital ecosystem where power is not concentrated in a few tech giants.

Appendix

Why is it so hard to get the message across?

1. **Profit-Driven Economy**: The current economic system is largely profit-driven, with many businesses depending on selling more products to grow. This often incentivizes consumption.

2. **Consumerism Culture**: Many societies value consumerism, often equating material wealth and consumption with success and happiness.

3. **Social Media Influence**: Platforms like Instagram and Facebook often promote a lifestyle of consumption and extravagance, influencing people to buy more.

4. **Fast Fashion and Planned Obsolescence**: Some industries, like fashion and electronics, promote frequent purchasing cycles with constantly changing trends and products designed to become obsolete quickly.

5. **Misinformation and Lack of Awareness**: Some people may not be fully aware of the impact of their consumption habits on the environment, or they may have been misled by misinformation.

6. **Convenience Culture**: The modern lifestyle often prioritizes convenience, which can lead to high energy consumption and waste.

7. **Perceived Deprivation**: Many people equate reducing consumption with sacrificing comfort, luxury, or quality of life.

8. **Lack of Alternatives**: In some cases, sustainable or low-consumption alternatives to common products or services may not be readily available or affordable.

9. **Inadequate Legislation and Regulation**: Government policies and regulations may not adequately incentivize or enforce sustainable consumption.

10. **Advertising Pressure**: Businesses spend billions on advertising campaigns that encourage people to buy more products or use more services.

11. **Peer Pressure and Social Norms**: People are often influenced by the consumption habits of their peers and may feel social pressure to consume at similar levels.

12. **Global Economic Inequality**: In some low-income regions, the focus is necessarily on meeting basic needs rather than reducing consumption.

13. **Growth-Oriented Policies**: Many political and economic policies emphasize growth as a key indicator of success, often equating it with increased production and consumption.

14. **Limited Scope of Individual Action**: Some people may feel that their individual efforts to reduce consumption won't make a difference in the grand scheme of things, leading to inaction.

15. **Psychological Factors**: Habits are hard to change, and cognitive biases like the status quo bias or the

immediacy bias can make people resistant to changing their consumption habits.

United Nations' 17 Sustainability Goals

Goals	Objective	Description
Goal -1	No Poverty	By 2030, eradicate extreme poverty for all people everywhere.
Goal -2	Zero Hunger	End hunger, achieve food security and improved nutrition by 2030.
Goal -3	Good Health and Well-being	Ensure healthy lives and promote well-being for all at all ages by 2030.
Goal -4	Quality Education	Ensure that all girls and boys complete free, equitable and quality primary and secondary education by 2030.
Goal -5	Gender Equality	To achieve gender equality and empower all women and girls.
Goal -6	Clean Water and Sanitation	Ensure availability and sustainable management of water and sanitation for all by 2030.
Goal -7	Affordable and Clean Energy	Ensure access to affordable, reliable, sustainable and modern energy for all by 2030.

Goal -8	Decent Work and Economic Growth	Promote sustained, inclusive and sustainable economic growth.
Goal -9	Industry, Innovation and Infrastructure	Build resilient infrastructure, promote inclusive and sustainable industrialization and foster innovation by 2030.
Goal - 10	Reduced Inequality	Reduce inequality within and among countries by 2030.
Goal - 11	Sustainable Cities and Communities	Make cities and human settlements inclusive, safe, resilient and sustainable.
Goal - 12	Responsible Consumption and Production	Ensure sustainable consumption and production patterns.
Goal - 13	Climate Action	Take urgent action to combat climate change and its impacts.
Goal - 14	Life Below Water	Conserve and sustainably use the oceans, seas and marine resources for sustainable development.
Goal - 15	Life on Land	Protect, restore and promote sustainable use of terrestrial ecosystems, combat desertification and halt biodiversity loss.
Goal - 16	Peace and Justice Strong Institutions	Promote peaceful and inclusive societies for sustainable development; provide access to justice for all.

Goal - 17	Partnerships to achieve the Goal	Strengthen the means of implementation and revitalize the global partnership for sustainable development.

Overlap between the great acceleration and the 17 UN SDGs

The "Great Acceleration" is a term for the rapid advancement in human activity from the mid-20th century onwards, leading to significant global environmental change. These problems often overlap with the United Nations' 17 Sustainable Development Goals (SDGs), which are designed to address a range of global challenges, including poverty, inequality, climate change, environmental degradation, and peace and justice.

Here are some ways these two concepts interact:

1. **SDG 1 (No Poverty)** and **SDG 2 (Zero Hunger)**: Great Acceleration impacts agricultural yields and food security, contributing to poverty and hunger.

2. **SDG 3 (Good Health and Well-being)**: Pollution and environmental degradation caused by the Great Acceleration negatively impact health and well-being.

3. **SDG 4 (Quality Education)**: The rapid pace of change can make educational content outdated quickly, which challenges the achievement of quality education.

4. **SDG 5 (Gender Equality)**: The impacts of environmental change often disproportionately affect women, who are often more vulnerable in times of crisis.

5. **SDG 6 (Clean Water and Sanitation)**: The Great Acceleration has led to water pollution and the overuse of water resources, making clean water and sanitation less accessible.

6. **SDG 7 (Affordable and Clean Energy)**: Accelerated energy consumption contributes to climate change and can limit access to affordable, clean energy.

7. **SDG 8 (Decent Work and Economic Growth)**: Rapid technological advancement can lead to job displacement, while environmental degradation can affect industries like agriculture and fishing.

8. **SDG 9 (Industry, Innovation, and Infrastructure)**: The Great Acceleration drives innovation but can also lead to infrastructural strain and increase the digital divide.

9. **SDG 10 (Reduced Inequalities)**: Environmental changes can exacerbate social and economic inequalities.

10. **SDG 11 (Sustainable Cities and Communities)**: Urbanization and rapid industrial growth can lead to unsustainable cities and communities.

11. **SDG 12 (Responsible Consumption and Production)**: The Great Acceleration has led to overconsumption and waste generation, challenging sustainable consumption and production.

12. **SDG 13 (Climate Action)**: The Great Acceleration has contributed significantly to anthropogenic climate change.

13. **SDG 14 (Life Below Water)** and **SDG 15 (Life on Land)**: Increased pollution, resource extraction, and

habitat destruction negatively impact biodiversity on land and in the oceans.

14. **SDG 16 (Peace, Justice, and Strong Institutions)**: Rapid change and environmental degradation can lead to social unrest and challenge the stability of institutions.

15. **SDG 17 (Partnerships for the Goals)**: The global nature of the impacts from the Great Acceleration necessitates international cooperation and partnerships to effectively tackle the challenges.

So, the Great Acceleration's problems interconnect with the SDGs in complex ways. Addressing the SDGs will often mean dealing with the issues stemming from the Great Acceleration, and vice versa.

Why such so slow progress on the UN's 17 SDGs?

1. **Lack of Resources**: Some governments, especially those in developing nations, may lack the financial, human, and infrastructural resources needed to implement and monitor the SDGs effectively.

2. **Political Instability**: In countries with unstable political climates, implementing long-term strategies such as the SDGs can be challenging.

3. **Short-term Focus**: Politicians often focus on short-term goals that align with their election cycles, while the SDGs require long-term commitment.

4. **Conflicting Interests**: Some SDGs may conflict with national interests, especially in countries heavily reliant on industries that contribute to environmental degradation or inequality.

5. **Inadequate Data and Monitoring Systems**: Many countries lack the data and systems needed to measure progress toward the SDGs accurately.

6. **Sovereignty Concerns**: Some governments may view the SDGs as an infringement on their sovereignty or a form of external influence or control.

7. **Lack of Public Awareness**: Without sufficient public understanding and support for the SDGs, governments may feel little pressure to implement them.

8. **Bureaucratic Hurdles**: Implementing the SDGs involves many sectors and levels of government, which

can be hindered by bureaucracy, poor coordination, or corruption.

9. **Prioritizing Economic Growth**: Some governments may prioritize economic growth over sustainable development, especially if they perceive a conflict between the two.

10. **Geopolitical Challenges**: Geopolitical issues, including conflicts, can divert attention and resources away from the SDGs.

11. **Inadequate Technological Infrastructure**: The lack of advanced technology or knowledge about implementing it can hinder the achievement of several SDGs.

12. **Resistance to Change**: Status quo bias or resistance to change in certain sectors could slow down or impede the implementation of SDGs.

13. **Insufficient International Cooperation**: Some SDGs require global efforts and cooperation, which can be challenging given geopolitical tensions and nationalistic tendencies.

14. **Limited Capacity for Implementation**: Governments may lack the institutional or policy-making capacity to fully implement the SDGs.

15. **Lack of Private Sector Engagement**: Successful implementation of SDGs often requires active participation from the private sector, which might not be present or willing in some contexts.

16. **Disparity in Regional Development**: Governments might struggle with uneven development within the

country, where some regions could be far behind others in achieving the SDGs.

17. **Shortage of Skilled Workforce**: Implementing SDGs effectively requires a skilled workforce in various sectors, which some nations might lack.

18. **Environmental Challenges**: Some countries, especially small island nations and those in environmentally vulnerable regions, face significant environmental hurdles that hinder SDG progress.

19. **Social Cultural Barriers**: In certain contexts, deep-rooted socio-cultural norms and attitudes may hinder progress on SDGs related to gender equality, education, and reduced inequalities.

20. **Political Will**: Ultimately, the successful implementation of the SDGs requires strong political will. Without it, even the most well-designed and well-resourced SDG strategies can falter.

While these are significant challenges, the SDGs offer a roadmap to a more sustainable, equitable, and prosperous world. Addressing these obstacles to their implementation is crucial, and will require creative thinking, international cooperation, and a commitment to long-term sustainability over short-term gains.

Barriers to achieving global wealth equality

1. **Unregulated Capitalism**: Unchecked capitalism can lead to wealth concentrating in the hands of a few, increasing wealth inequality.

2. **Corruption**: Corruption in political and economic systems can hinder the fair distribution of wealth and perpetuate inequality.

3. **Inadequate Education**: Lack of access to quality education, particularly in low-income areas, can limit people's ability to improve their economic situation, exacerbating wealth inequality.

4. **Tax Evasion and Avoidance**: Wealthy individuals and large corporations often exploit legal loopholes to avoid paying their fair share of taxes, contributing to wealth inequality.

5. **Political Influence**: Wealthy individuals and corporations often have disproportionate influence on political systems, allowing them to shape policies in ways that perpetuate wealth inequality.

6. **Discrimination**: Systemic discrimination based on race, gender, sexual orientation, and other factors can limit economic opportunities for marginalized groups and contribute to wealth inequality.

7. **Limited Social Mobility**: In many societies, it's difficult for individuals to move up the economic ladder due to factors like poor education, lack of opportunities, and entrenched class systems.

8. **Inadequate Labor Laws**: Weak labour laws can lead to low wages, poor working conditions, and job insecurity, particularly for workers in low-income jobs, further exacerbating wealth inequality.

9. **Lack of Financial Literacy**: Without knowledge about managing finances, making investments, and navigating financial systems, individuals may struggle to build wealth.

10. **Globalization**: While globalization has many benefits, it can also contribute to wealth inequality by enabling corporations to exploit cheap labour in developing countries, driving down wages, and by fostering competition between nations that can lead to wage suppression.

These barriers are substantial and deeply rooted, but they are not insurmountable. Addressing them will require significant efforts at various levels, from local communities to global institutions, and will involve measures such as policy reforms, educational initiatives, and systemic changes to economic and political systems.

How can we decouple GDP from quality of life?

Decoupling global GDP from personal quality of life involves redefining success beyond pure economic growth to incorporate aspects of well-being, social equity, and environmental sustainability. Here are some potential strategies:

1. **Redefining Measures of Success**: Adopt measures such as the Genuine Progress Indicator (GPI) or Gross National Happiness (GNH) that incorporate social and environmental factors in addition to economic output.

2. **Universal Basic Income (UBI)**: Implementing a UBI can provide everyone with a base level of financial security, improving quality of life regardless of economic growth.

3. **Access to Quality Healthcare**: Ensure universal access to quality healthcare, as health is a fundamental component of quality of life.

4. **Education**: Prioritize access to quality education at all levels, which can improve quality of life and promote sustainable development.

5. **Social Safety Nets**: Expand social safety nets to protect individuals from economic downturns and ensure a minimum standard of living.

6. **Sustainable Development**: Transition to models of sustainable development that prioritize environmental and social well-being in addition to economic growth.

7. **Income Equality**: Implement policies to reduce income and wealth disparities, as excessive inequality can have negative impacts on societal well-being.

8. **Work-Life Balance**: Encourage work-life balance through policies like flexible working hours, paid time off, and parental leave.

9. **Affordable Housing**: Ensure access to affordable, safe, and adequate housing, a basic need tied closely to quality of life.

10. **Community Development**: Foster strong, vibrant communities and invest in public goods like parks, libraries, and public transportation.

11. **Access to Healthy Food**: Support systems for providing access to nutritious, affordable food, as diet significantly impacts health and well-being.

12. **Mental Health Services**: Prioritize mental health services alongside physical health, with accessible and stigma-free support systems.

13. **Public Participation**: Encourage public participation in decision-making processes, which can increase societal buy-in and satisfaction.

14. **Cultural and Recreational Opportunities**: Invest in arts, culture, and recreational activities that contribute to happiness and personal enrichment.

15. **Clean Environment**: Policies to maintain clean air, water, and other environmental resources, as a healthy environment is vital for human well-being.

These solutions shift focus from an almost exclusive concentration on economic output to a more balanced approach that considers other critical aspects of human well-being. Implementing these changes requires political will, policy shifts, and societal support.

What does it mean to be a human being?

The definition of fulfilment varies from person to person, largely based on cultural, personal, and societal values. However, here are 15 possible elements that could contribute to a sense of being a fulfilled human being:

1. **Purpose**: Having a sense of purpose, direction or meaning in life, often derived from work, hobbies, or contributions to community.

2. **Connection**: Developing deep and meaningful relationships with other people.

3. **Self-Expression**: Feeling free to express oneself and one's individuality.

4. **Achievement**: Attaining goals and experiencing a sense of accomplishment.

5. **Growth**: Continually learning, evolving, and expanding one's skills, knowledge, or understanding.

6. **Self-Acceptance**: Accepting oneself, including both strengths and weaknesses, cultivating self-love and compassion.

7. **Health**: Enjoying physical health and vitality, which can contribute to mental and emotional well-being.

8. **Contribution**: Making a positive difference in the lives of others or the world at large.

9. **Resilience**: The ability to cope with life's ups and downs, and to bounce back from adversity.

10. **Contentment**: Feeling a sense of satisfaction with life as it is, rather than constantly striving for more or different.

11. **Creativity**: Engaging in creative activities or creative problem-solving.

12. **Mindfulness**: Being present and engaged in the current moment, rather than dwelling on the past or worrying about the future.

13. **Joy**: Experiencing joy and happiness, from daily pleasures or deeper sources.

14. **Inner Peace**: Experiencing a sense of inner calm and tranquillity, often through practices such as meditation or spending time in nature.

15. **Autonomy**: Having the freedom and ability to make choices and decisions for oneself.

Remember, what brings fulfilment can vary greatly from person to person, and it's a deeply personal journey to discover what makes one feel truly fulfilled. It can be helpful to reflect on what feels meaningful and satisfying, and to seek experiences and relationships that align with those values.

Personal carbon dioxide emission rationing

The exact value for personal carbon dioxide emissions that would keep global warming under 1.5 degrees Celsius is complex to estimate and varies depending on many factors. However, to give a ballpark figure, let's use the global carbon budget.

The Intergovernmental Panel on Climate Change (IPCC) in their 2018 special report estimated that the remaining carbon budget to keep warming under 1.5 degrees Celsius with a 66% chance was about 420 gigatons of CO2 at the start of 2018. Considering that the world population is around 7.8 billion as of 2020 and is expected to reach 8.5 billion by 2030, let's use an average of about 8.15 billion people.

If we were to divide the 420 gigatons of CO2 by 8.15 billion people, we'd get roughly 51.5 tons of CO2 per person from 2018 onwards.

However, it's also important to note that global emissions are currently about 42 gigatons per year (as of 2020). At this rate, the carbon budget for 1.5 degrees Celsius would be used up in about 10 years.

These figures should only be used as a rough estimate because they do not take into account many other factors, such as the fact that some level of carbon emissions is almost unavoidable in our current society, the different emission levels between countries and individuals, and the impact of other greenhouse gases.

So in an equal world scenario, if everyone was assigned an equal share of the remaining carbon budget, the amount would be quite low and would require drastic changes in our lifestyles and energy systems to achieve. This emphasizes the urgency of taking action on climate change.

The annual per person carbon dioxide emissions allowance can be calculated by taking the overall personal allowance and dividing it by the number of years.

We had an estimated 51.5 tons (or 51,500 kilograms) of CO_2 per person from 2018 onwards to stay below 1.5 degrees Celsius of warming. If we were to divide this by, say, 10 years (considering the urgency of climate change and the rapid depletion of the carbon budget), we'd get roughly 5,150 kilograms (or 5.15 tons) of CO_2 per person per year. However, this calculation is a very simplified version of a much more complex problem. It doesn't take into account a range of factors such as:

1. The global carbon budget is being rapidly depleted, and at current rates of emissions, the budget for 1.5 degrees Celsius would be used up in less than 10 years.

2. There's a significant variation in per-capita emissions among different countries. For instance, the average American has a much larger carbon footprint than the average resident of a developing country.

3. This calculation also assumes a linear decrease in emissions, but in reality, transitioning to low-carbon technologies is likely to involve a more rapid decrease in emissions in the later part of the transition.

So while the figure of approximately 5,150 kg of CO_2 per person per year gives an idea of the magnitude of the challenge, it doesn't fully capture the complexities of the issue.

Personal Polycrisis Action Plan

Creating a sustainable domestic living environment involves numerous changes to lifestyle, consumption, and technology. The annual carbon dioxide emissions for each solution can be difficult to quantify precisely, as it depends on various factors like household size, local climate, individual lifestyle choices, and the carbon intensity of the local electricity grid. Nevertheless, I can provide a rough estimate based on available data up to my knowledge cut-off in September 2021. Here are the top 15 solutions, ordered from most to least impactful:

1. **Shift to Renewable Energy**: If a household switches to renewable energy sources like solar or wind power, it can significantly cut carbon emissions. This could potentially save thousands of kilograms of CO_2 emissions each year, depending on the household's energy usage.

2. **Electric Vehicles (EVs)**: If a household switches from a gasoline car to an electric vehicle and charges it with renewable energy, this could save approximately 2,000 - 3,000 kg of CO_2 per year.

3. **Energy Efficiency**: Improving home energy efficiency through better insulation, energy-efficient appliances, and smart thermostats can lead to significant reductions. This could save around 500 - 1,000 kg of CO_2 per year.

4. **Low Carbon Diet**: Switching to a plant-based diet can significantly reduce a household's carbon footprint, potentially saving around 1,000 kg of CO_2 per person per year.

5. **Reduce, Reuse, Recycle**: Following the three R's (reduce, reuse, recycle) can lead to significant carbon savings by reducing the energy used to produce new goods. This could save roughly 500 - 1,000 kg of CO_2 per year.

6. **Public Transport/Cycling/Walking**: Using public transport, cycling, or walking instead of personal cars for commuting can save around 500 - 1,000 kg of CO_2 per person per year.

7. **Water Conservation**: Implementing water-saving strategies can also save energy, leading to a reduction of around 200 - 300 kg of CO_2 per year.

8. **Waste Composting**: Composting organic waste can save around 50 - 100 kg of CO_2 per year by reducing methane emissions from landfills.

9. **Smart Lighting**: Switching to LED bulbs and using smart lighting systems can save around 50 - 100 kg of CO_2 per year.

10. **Green Gardening**: Sustainable gardening practices, like composting and capturing rainwater, can lead to small carbon savings and other environmental benefits.

11. **Local and Seasonal Food**: Buying local and seasonal food can reduce the carbon footprint associated with transporting food, potentially saving around 50 - 100 kg of CO_2 per year.

12. **Eco-Friendly Products**: Using eco-friendly products (cleaning products, personal care products, etc.) can reduce the indirect carbon footprint of a household.

13. **Clothes Line Drying**: Avoiding the use of electric dryers by line drying clothes can save around 50 - 100 kg of CO2 per year.

14. **Mindful Shopping**: Buying only what is needed and opting for second-hand or sustainably produced items can lead to modest carbon savings.

15. **Tree Planting**: Planting trees can help absorb CO2, although the impact in terms of annual emissions reduction is small in the early years of a tree's life.

Note: These figures are rough estimates and actual carbon savings could vary widely depending on numerous factors. It's also important to remember that all these actions can contribute towards sustainability, and they often have additional environmental benefits beyond reducing carbon emissions.

Small Business Polycrisis Plan

Running a sustainable small business involves numerous changes to your operations, consumption, and technology. The annual carbon dioxide emissions for each solution can be hard to quantify precisely, as it depends on various factors like business size, industry, location, and specific practices. However, here are the top 15 solutions, ordered from most to least impactful, with rough estimates based on available data up to my last training cut-off in September 2021:

1. **Switching to Renewable Energy**: Switching your business operations to renewable energy sources can significantly reduce carbon emissions. Depending on the size and nature of your business, this could save many thousands of kilograms of CO_2 emissions each year.

2. **Energy Efficiency**: Improving energy efficiency in your business premises by using energy-efficient appliances, better insulation, and smart technologies can significantly cut emissions. This could save around 1,000 - 5,000 kg of CO_2 per year, depending on your business size and current practices.

3. **Sustainable Transportation**: Encouraging the use of public transportation, carpooling, cycling, or walking for commuting or business travel can save around 500 - 2,000 kg of CO_2 per employee per year.

4. **Digitalization**: Digitalizing processes and reducing paper usage can save approximately 100 - 500 kg of CO_2 per year.

5. **Recycling and Waste Reduction**: Implementing comprehensive recycling programs and reducing waste

in your operations could save around 200 - 1,000 kg of CO2 per year.

6. **Sustainable Procurement**: Procuring goods and services from sustainable sources can reduce the indirect carbon footprint of your business, potentially saving 100 - 500 kg of CO2 per year.

7. **Telecommuting**: Allowing employees to work from home can reduce commuting emissions, saving around 100 - 500 kg of CO2 per employee per year.

8. **Green Web Hosting**: Choosing a web hosting service that operates on renewable energy can save around 50 - 100 kg of CO2 per year.

9. **Offsetting Emissions**: Investing in carbon offset projects can compensate for emissions that can't be eliminated, although it's more beneficial to reduce emissions directly where possible.

10. **Plant-Based Catering**: Offering plant-based food options at business events can reduce the carbon footprint associated with food, saving around 10 - 50 kg of CO2 per event.

11. **Green Marketing**: Promoting your business's sustainability efforts can encourage other businesses and customers to make more sustainable choices.

12. **Local Sourcing**: Sourcing goods and services locally can reduce the carbon footprint associated with transport, saving around 10 - 50 kg of CO2 per year.

13. **Sustainable Packaging**: Using minimal, recyclable, or compostable packaging can reduce the carbon footprint

associated with packaging, saving around 10 - 50 kg of CO_2 per year.

14. **Water Conservation**: Implementing water-saving measures in your business premises can save energy and reduce carbon emissions, although the impact is likely to be small for most businesses.

15. **Education and Training**: Educating employees about sustainability and how to reduce their carbon footprint can lead to further carbon savings, although it's difficult to quantify the impact.

Remember, these figures are rough estimates and actual carbon savings could vary widely depending on many factors. Every business is unique and should assess its own sustainability challenges and opportunities. A detailed carbon audit can provide more precise information about where a business can reduce its emissions.

Printed in Great Britain
by Amazon

30432164R00069